W9-BNT-584

Eyes on the Sky

Saturn

by David M. Haugen

MAY 1 3 2003

CARTER PUBLIC LIBRARY

KidHaven Press

KidHaven Press, an imprint of Gale Group, Inc.

10911 Technology Place, San Diego, CA 92127

Library of Congress Cataloging-in-Publication Data
Haugen, David M., 1969–
    Saturn/by David M. Haugen
        p. cm.—(Eyes on the Sky)
    Includes bibliographical references and index.
    Summary: Discusses Saturn, its makeup, its moons, the
    composition and formation of its rings, and scientific obser-
    vations of the planet through space missions and the use of
    telescopes.
    ISBN 0-7377-0758-5
    1. Saturn (Planet)—Juvenile literature. [1. Saturn (Planet)]
    I. Title.
    QB671.H22 2002
    523.46—dc21

    00-012942

Picture Credits

Cover photo: PictureQuest
Associated Press/NASA, 11
© CORBIS, 37 (bottom)
© J. Calvin Hamilton/Jet Propulsion Laboratory, 26
JPL/NASA, 10, 12, 15, 28, 31 (right), 34 (top), 40
Chris Jouan, 24, 32–33
NASA, 7, 17, 31 (left), 38
© NASA/Roger Ressmeyer/CORBIS, 23, 34 (bottom), 37 (top)
PictureQuest, 1
© Premium Stock/CORBIS, 22
Scala/Art Resource, 6
Martha Schierholz, 9, 19

Copyright 2002 by KidHaven Press, an imprint of Gale Group, Inc.
        10911 Technology Place, San Diego, CA 92127

No part of this book may be reproduced or used in any other
form or by any other means, electrical, mechanical, or otherwise, including, but not limited to, photocopying, recording, or
any information storage and retrieval system, without prior
written permission from the publisher.

Printed in the U.S.A.

Table of Contents

1
Observing the Ringed Planet

The planet Saturn with its prominent rings has fascinated stargazers for centuries. Although the earliest observers could only view Saturn as a faint light in the nighttime sky, it still drew their attention because it appeared unlike the stars that surrounded it. Ancient peoples recognized that Saturn does not twinkle like the stars. They also noticed that it has an unusual way of moving across the sky throughout the year. While stars trace a steady path across the heavens, Saturn seems to occasionally shift from side to side as it follows its course. The ancient Greeks noticed this odd movement and referred to Saturn as a planet— meaning "wanderer"—to emphasize that it was different from the stars.

Looking into the sky with the naked eye was the only method ancient civilizations had to observe Saturn. They could note the unusual behaviors of the planet, but they could not explain them. For example, ancient astronomers had no way of knowing that Saturn's shifting motion was due to its circular movement, or **orbit**, around the sun. Likewise, they didn't recognize that Earth was also a planet orbiting the sun. Today, however, astronomers realize that the sun is the central point around which all the planets in this **solar system** revolve. Currently astronomers believe there are nine planets in the solar system. All of these planets orbit the sun at different distances. Earth is the third planet in line, orbiting at about 93 million miles from the sun. Saturn is the sixth planet, orbiting at almost 890 million miles from the sun.

The Telescope Reveals Rings

Distance and orbit, however, were facts uncovered by telescopes, instruments developed in the seventeenth century. The earliest telescopes were rather crude and could only help to slightly enlarge the image of heavenly bodies. In Saturn's case, however, the first telescopes greatly increased what astronomers knew about the planet. In 1610 the Italian astronomer Galileo Galilei used a telescope to

Galileo saw Saturn's rings through his telescope.

view Saturn. To his surprise, the pale yellow ball appeared to have two bulges extending from its sides. Galileo assumed the bulges were really two other smaller planets that traveled in a group with the main body. The Italian astronomer concluded that Saturn was not one planet but three.

Galileo continued to watch Saturn. Within two years he noticed that the bulges he believed to be planets disappeared and then strangely reappeared a few years later. These observations made him abandon his three-planet theory, but he was at a loss to explain Saturn's bulges. Not until a Dutch scientist named Christiaan Huygens viewed the planet in 1655 did a new theory emerge. Using an improved telescope, Huygens saw that the bulges were not planets, but rather a set of rings that circled the planet. The rings seemed to disappear over time because they were set at an angle to Saturn's main globe. Since Saturn revolves as it moves across the sky, the thin rings seem invisible when they are level to the path of observation on Earth.

Improving Knowledge of Saturn

As telescopes improved over the centuries, so did astronomers' knowledge of the ringed planet. By combining measurements of the

Saturn's rings cannot always be seen from Earth.

visible surface of Saturn with simple geometry, astronomers were able to determine that Saturn was 74,800 miles wide at its equator. That's almost ten times as wide as Earth. In fact, if Saturn were hollow, it could hold 760 Earths. This made Saturn the second largest planet in the solar system. Only the huge planet Jupiter was bigger.

Sharper images from telescopes with greater magnification also showed that Saturn was not a uniformly yellow ball. The visible surface of the planet is made up of broad, pale cloud bands. The cloud bands run parallel to Saturn's equator. Each is hundreds of miles wide, and the bands near the equator are much wider than the bands near the planet's north and south poles. The bands are formed by wind patterns. The winds in each band travel at varying speeds. The winds near the equator can gust up to 1,100 miles per hour while those near the poles move slower.

While watching the movement of the cloud bands through their telescopes, scientists learned something new about Saturn. They learned that Saturn rotates at very high speeds. Unlike Earth's 24-hour day, Saturn spins once every 10 hours and 39 minutes. Because the planet's surface is not solid, different parts of Saturn spin at different speeds. The cloud bands near the equator complete one

Saturn's Place in the Solar System

Pluto
Neptune
Uranus
Saturn
Jupiter
Mars
Earth
Venus
Mercury
Sun

The ringed planet is a gas giant.

rotation in 10 hours and 10 minutes, roughly a half hour before the bands at the poles bring Saturn's day to a close. The rapid rotation causes the planet to bulge at the equator, making Saturn look like a slightly squashed ball.

Revealing a Gas Giant

Optical telescopes, those that provide information based on what scientists can see through the eyepiece, added to the growing pool of information about the planet. They could show size, shape, color, and movement. These telescopes, however, revealed nothing about what lay beneath the clouds of the yellowish planet.

In the 1950s telescopes that use radio signals to help gain more accurate readings of the distant planets uncovered something more about Saturn. These radio telescopes bounce radio waves off distant objects and interpret the signals as they return to Earth. The radio signals showed that Saturn was not a solid body; instead the huge planet seems to be made up entirely of gases. Because of this, Saturn is called a gas giant.

Viewing Saturn from Space

Scientists wanted to learn more about Saturn than their earthbound telescopes could tell them. This new knowledge finally came in the 1970s when orbiting space probes and telescopes provided the first close-up views of Saturn.

The Hubble Space Telescope shows a large whirlpool storm on Saturn.

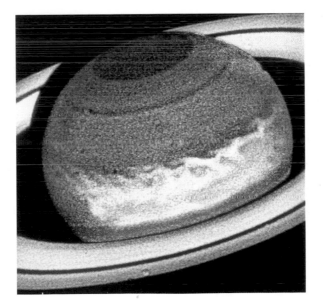

For example, in 1990 the Hubble Space Telescope that orbits Earth noticed a large white oval near Saturn's equator. Called the Great White Spot, the oval shape is a whirlpool storm as big as Earth. Another, smaller oval

Voyager 1 (pictured) gave scientists new information about Saturn.

was spotted nearby in 1994. These storms are similar to the type that arise on the planet Jupiter, but on that planet the whirlpools' colors are so bright that they easily stand apart from other cloud formations. On Saturn the white shapes are nearly invisible against the yellowish color of its surface. They also seem to arise and fade over time. Not enough information has yet been gathered to determine whether these spots return on a regular basis.

Providing Valuable Information

Like the pictures from the Hubble Telescope, three space probes sent to Saturn have also provided valuable images and information about the planet that astronomers could otherwise not obtain. *Pioneer 11*, *Voyager 1*, and *Voyager 2*—all launched in the 1970s by the National Aeronautics and Space Administration (**NASA**)—have traveled by the distant planet, sending back data as they have passed. Although none of these probes entered the planet's atmosphere, instruments on board were able to take readings that allow scientists to make more educated guesses about Saturn's interior.

2
Inside Saturn

In 1980 and 1981, when *Voyager 1* and *Voyager 2* flew by Saturn, data from these space probes revealed much about the interior of the planet. These same probes had already passed by Jupiter, and astronomers believed that the two gas planets had many similar features. The Saturn flyby confirmed this belief, yet enough differences were noted for scientists to realize that Saturn was not just a smaller version of its huge neighbor.

A Hydrogen-Rich Planet

Like Jupiter, Saturn is mostly made up of **hydrogen**. Saturn's hydrogen exists mainly as a gas in the atmosphere. Beneath the clouds of gas, however, where temperatures increase, scientists believe the hydrogen becomes a liquid. In fact, they think a vast ocean of liquid hydrogen may surround the planet's core.

Saturn has more hydrogen than any other planet. More than 91 percent of Saturn is made up of hydrogen. Another 6 percent of the

Hydrogen gas makes up much of the atmosphere of Saturn (left) and Jupiter (right).

planet is **helium** gas. The abundance of these two gases means the huge planet is light for its size. The entire planet is less dense than water on Earth, so if an ocean of water large enough could be found, Saturn would float atop it.

Scientists once assumed that Saturn was similar in composition to the sun, which also contains a lot of hydrogen and helium. In fact, some scientists believed that Saturn almost became a star when the solar system was forming. Recent findings, however, show that Saturn has an even greater percentage of hydrogen than the sun, so conditions were probably never just right for Saturn to become a star.

Other Elements and Compounds

Other gases detected in Saturn's atmosphere include **methane** and **ammonia**. Since no probe has descended into Saturn's atmosphere, scientists can only guess at what other substances might be present within the planet. Common hydrogen compounds like water and hydrogen sulfide are suspected to exist in deeper levels of the atmosphere but so far have not been detected.

Although the gases that make up the majority of Saturn's atmosphere are colorless,

Saturn appears pale yellow because a layer of haze covers its outer surface.

Saturn's clouds display a wide range of colors. These colors are not noticeable in most images taken from space or from Earth because a layer of haze covers the outer surface, muting all the colors to pale yellow. If the haze were peeled away, the cloud bands underneath would appear in much more vivid yellow, brown, and red hues. Astronomers think the colors are caused by traces of **phosphorus** mixed in with the clouds' gases.

Temperatures

The upper cloud layers are also the coldest places on the planet. Here temperatures are more than 300 degrees Fahrenheit below zero. These regions are colder than any place on Earth.

Although the atmosphere of the planet is very frigid, the area around the planet's center is extremely hot. Temperatures there rise to over 11,000 degrees Fahrenheit.

The intense heat of the planet's core is something of a mystery. Some scientists think that matter near the center of the planet is being compressed so much by the planet's **gravity** that heat is generated. Another theory proposes that friction caused by molecules falling, or raining, toward the planet's center may produce this heat. Since no space probe could stand the crushing pressures of a descent to the center of the planet, however, there is no way to know for sure the cause of Saturn's internal heat.

Evidence of a Planetary Core

Even though no probe could reach Saturn's center, most astronomers believe the planet has a solid core. This belief results from observations of Saturn's gravitational field. Every planet has a gravitational field. On Earth that field draws all things toward the ground. Since Saturn is made up of gases, all matter is drawn toward the planet's center.

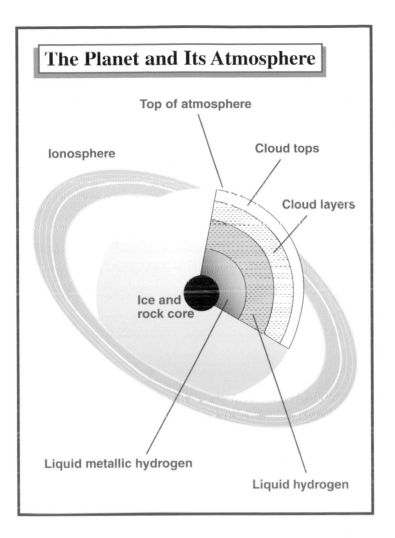

The Planet and Its Atmosphere

Top of atmosphere

Ionosphere

Cloud tops

Cloud layers

Ice and rock core

Liquid metallic hydrogen

Liquid hydrogen

Astronomers have measured Saturn's gravitational field by the way it attracted the spacecraft that have passed by the planet. This information revealed that the planet's gravitational field is not perfectly round as one would expect from a planet with no solid core. The field's odd shape tells scientists that elements heavier than hydrogen may be found at the center of the planet. This suggests that Saturn has a solid center, probably made of rock and ice.

Continuing Study

Astronomers expect that future space missions to Saturn will provide more information about the planet's interior. Scientists are eager to learn more about Saturn's atmosphere as well as the weather patterns of its skies. They also hope that new probes will send pictures and information about Saturn's moons and its rings. To astronomers, these two aspects of the planetary system are as important to examine as the great body of Saturn itself.

3
Saturn's Rings

The rings that encircle Saturn are its most striking and well-known feature. Astronomers have known about the rings since early telescopes were first trained on the planet. For centuries the rings made Saturn unique. In 1977, however, faint rings were also discovered around the planet Uranus, and within two years similar rings were found on both Jupiter and Neptune. Still, Saturn's rings remain the most impressive in the solar system, mainly because they are so large and easy to see.

From One Ring to a Thousand Ringlets

In 1655, when Dutch scientist Christiaan Huygens first determined that Saturn had a ring system, he assumed the entire flat disk that he saw with his telescope was a single, solid ring. Twenty years later the Italian astronomer Gian Domenico Cassini spotted a gap within the rings. Called the Cassini Division, this gap disproved Huygens's theory since two rings could be seen: an inner and an outer ring. Then, in 1837 a German astronomer named Johann Encke reported another gap, this time within the outer ring. The Encke Division was so hard

The Cassini Division, which shows as a dark gap, divides Saturn's rings.

Voyager 1 found more than one thousand different rings around Saturn.

to see, however, that its existence was not confirmed until 1978, when high-powered telescopes made it clearly visible.

Once earthbound observation was improved, astronomers began to notice that more rings and divisions existed. Scientists realized that the wide, flat rings were really many small rings, or ringlets, bunched together. When the *Voyager 1* spacecraft passed by the planet, it revealed more than a thousand different ringlets. Some of these were round; others were oval shaped. Some stood alone; others wound together. None were more than five hundred feet thick, making them hard to see from Earth.

Despite the knowledge that the broad, flat rings were really many ringlets, astronomers still group the ringlets into seven main rings. The rings are labeled A through G, though they are not ordered alphabetically. Only the A, B, and C rings are visible from Earth. Of these, the B ring, located toward the center of the ring system, is the easiest to see. The D, E, F, and G rings were only detected by passing space probes. Of these, the outermost, or E ring, of Saturn is so faint that it is hard to detect even from space.

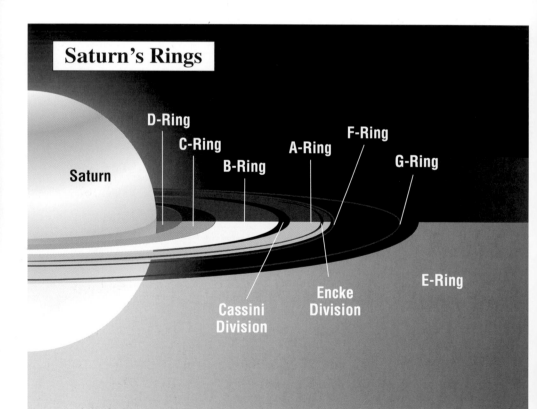

The Composition and Formation of the Rings

The rings are made up of millions of chunks of ice and probably some rock. Some of these pieces are smaller than a penny; others are bigger than a house. Together, these floating objects reflect sunlight, which is what makes the rings visible. The brightness of each ring is determined by how wide and how thick the ring is and how much space exists between the material within them. For example, even though the outer E ring measures 187,000 miles across, the pieces of ice and rock within it are scattered throughout the huge area, making the ring barely visible.

Astronomers are not sure where the material in Saturn's rings came from. One theory proposes that some of Saturn's large moons may have been blasted by **comets** or **meteors**, spreading the debris into space. Whatever the cause, the bits of ice and rock revolve around the planet because they are stuck in its orbit, attracted by Saturn's gravity but moving through space just fast enough not to "fall" toward the planet.

Forces related to the planet's gravity also keep the larger pieces of material from coming together. That is, the larger masses within the rings would normally attract each other because

CARVER PUBLIC LIBRARY

they have gravity of their own, but a force generated by the planet balances these gravitational forces. Therefore, the material cannot combine to form even larger masses that would disrupt the ring pattern, probably by sweeping away the smaller particles in the rings.

Moons and the Ring Gaps

Some scientists believe that the gaps between Saturn's rings are maintained by objects that are already of the size necessary to sweep away small debris from their paths. Some of Saturn's smaller moons—or **satellites**—are big enough to exert a force upon neighboring pieces of rock and ice in the rings. The tiny moon Pan, for example, was unknown until

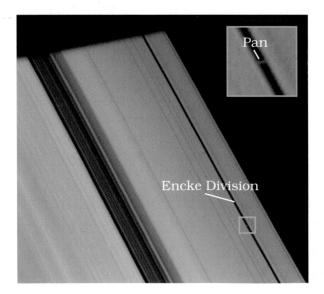

Inset shows a close-up view of the tiny moon Pan in the Encke Division.

the forces it was exerting on ring material indicated to astronomers that it was hidden within the Encke Division. Without something constantly clearing the paths in between the rings, astronomers know the gaps would fill in with matter from the surrounding rings.

Astronomers also recognize that not all the gaps are caused by moons simply bulldozing a path through the ring material. Instead, pairs of satellites—with one on each side of a ring—sometimes act as shepherds to keep rocks and ice within the boundaries of the rings and out of the gaps. Known as shepherd satellites, these tiny moons exert forces on the ring material that keep it from drifting. Many such shepherd satellites may exist within Saturn's ring system, but scientists are unsure of the exact number. Cameras on the *Voyager* spacecraft did not send back any images to help astronomers pinpoint the locations of the many smaller moons they believe are hidden among the debris within the planet's rings.

More Unsolved Mysteries

The Voyager missions did find evidence of another unusual feature of Saturn's rings that astronomers on Earth are unable to detect. Dark wedges within the B ring appear as "spokes" within the circular shape. Scientists think these

Scientists hope the *Cassini* spacecraft will send home new information about Saturn and its rings.

spokes are made of fine particles of rock or ice, but the *Voyager* craft did not get close enough to know for certain. Viewing images taken over time, astronomers do know the spokes seem to arise and then disappear. Still, no one is sure how or why the spokes form at all.

The spokes of Saturn's B ring will remain one of the unsolved mysteries of Saturn's ring system, at least until the *Cassini* spacecraft reaches the planet in the early twenty-first century. Only then, perhaps, will astronomers find answers to the many unsolved mysteries of the ringed planet.

4
Saturn's Moons

Although astronomers are eager to find out more about Saturn's interior and its ring system, they are also fascinated by the planet's moons. Saturn has more moons than any other planet in the solar system. So far eighteen have been positively identified, but other ones are likely to be found. Some of these moons are large bodies like Earth's moon; others are tiny satellites, rocks too small to be thought of as proper moons yet are often considered such. Many of these minor satellites have been seen in the Saturn system, but only a few have been confirmed by repeated sightings.

Floating Rocks

Most of Saturn's moons and satellites lie along the planet's equatorial plane; that is, they are located in an imaginary disc that extends from the equator of the planet—just as Saturn's rings do. The small satellite Pan resides within the main visible ring system. Pandora, Epimetheus, and Janus orbit just outside the visible rings, while Atlas and Prometheus act as shepherd satellites on either side of the faint F ring.

All of these satellites are very small, with only a very few measuring over a hundred miles across. They are most likely larger versions of the chunks of rock and ice that are trapped in Saturn's rings. Other minor satellites such as Telesto, Calypso, and Helene exist well outside of Saturn's visible rings. These, too, are odd-shaped rocks floating in space.

Mimas and Enceladus

Saturn's medium-sized moons orbit at least thirty thousand miles beyond the visible edge of Saturn's A ring. The closest is Mimas, a round moon measuring about 245 miles across its equator. The surface of Mimas is covered with craters, the ring-shaped ridges that indicate objects have crashed into the moon. Almost all of Saturn's moons show evidence of impact with meteors or other large space debris. Mimas, however, has the largest impact crater of any of the planet's moons.

Called the Herschel crater, this large region is about eighty miles across, or roughly a third of the moon's **diameter** (the distance across the equator of the moon). The crater is also about six miles deep, suggesting that a huge and heavy object must have struck the moon.

Enceladus, the next medium-sized moon from the planet, is a little larger than Mimas. Unlike the other moons, Enceladus shows relatively few large impact craters. Scientists are not sure why this is, but they suspect that Enceladus has experienced volcanic activity within the past 100 million years. Internal heat may have melted the ice on the moon's surface, allowing it to spread over older craters. When the moon cooled, the ice froze again and hid any evidence of past impacts.

Photographs show the many craters of Mimas (left) and far fewer craters on Enceladus.

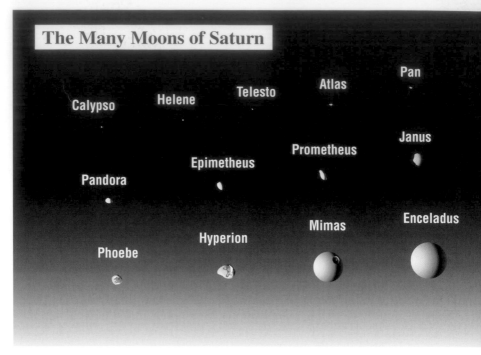

The Many Moons of Saturn

Pan
Atlas
Telesto
Calypso   Helene
Janus
Prometheus
Epimetheus
Pandora
Enceladus
Mimas
Hyperion
Phoebe

Source: Monterey Institute for Research in Astronomy.

Tethys, Dione, and Rhea

Saturn's larger inner moons are Tethys, Dione, and Rhea. Tethys shows no signs of recent activity from internal heat. It's surface is heavily cratered, indicating that the moon's icy covering has not moved in a long time. Tethys also has a large valley that runs from one side of the planet, over the north pole, and down the other side. Some scientists think that the valley may have been caused by the impact of a large rock from space. Others believe it was created when water just under the surface of the newly formed moon froze and then cracked.

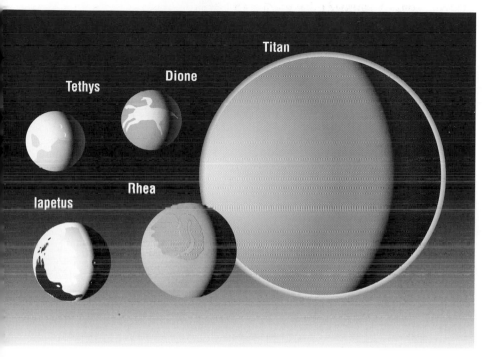

Like Tethys, the moons Dione and Rhea
are also covered with craters. Rhea, in fact,
has more craters on its surface than any of
Saturn's other moons. This suggests that
Dione has probably experienced movement of
its icy surface more recently than Rhea,
though still millions of years ago. Scientists
find this unusual because Rhea is the larger
of the two moons, and its internal temperatures should have taken longer to cool. With a
diameter of about seven hundred miles,
Dione is just slightly bigger than Tethys. Rhea
is the second largest of Saturn's moons,
measuring about 950 miles across.

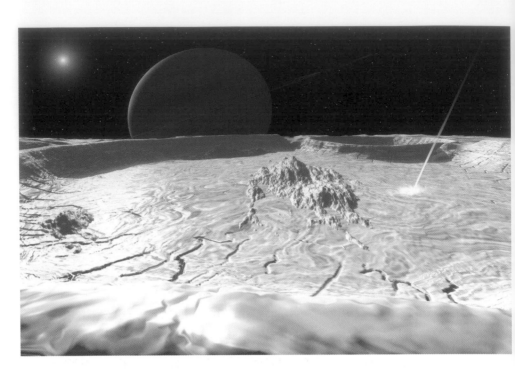

Rhea's surface is covered with craters.

Saturn dwarfs two of its many moons, Tethys and Dione.

Titan

Saturn's largest moon is named Titan. It measures more than three thousand miles across, making it even larger than the planet Mercury. Titan is an unusual moon because it has a dense atmosphere. A few other moons in the solar system have atmospheres, but not as thick or deep as Titan's.

Astronomers are interested in this unique moon, so the *Voyager 1* space probe was sent very close to Titan to see if more could be learned about its atmosphere. *Voyager*'s cameras were not able to see past a heavy brownish haze that covers the moon, but the probe's other sensors were able to send back a lot of information about the moon and its atmosphere. Titan's atmosphere consists mostly of nitrogen with small amounts of methane and carbon dioxide.

The large amount of nitrogen suggests Titan's atmosphere is similar to that of Earth. Because of this, some scientists think that some very early building blocks of life may have once existed on Titan. These traces of life-generating compounds could not form into a living organism, however, because the temperatures on the moon are too cold to support life. Temperatures on Titan's frigid surface reach 294 degrees Fahrenheit below zero. Scientists hope to learn much more about this interesting moon when the *Cassini* spacecraft launches its probe into Titan's atmosphere.

The Outermost Moons

While the *Voyager 1* spacecraft helped reveal much about Titan, no space probe has even come near Saturn's three outermost moons. Hyperion, Iapetus, and Phoebe are so far from the planet and the course of the Voyager missions that few good photographs of these moons exist.

Hyperion is a large, oblong moon, about 250 miles long and 150 miles wide. Its irregular, rocky shape resembles the minor satellites that orbit near Saturn's rings. However, the moon is too large to be related to those floating rocks, and astronomers suspect Hyperion has not always looked so jagged. Some believe Hyperion was once a larger, round moon, but that it was blasted apart by a large meteor that crashed into it long ago.

Iapetus is another of the unexplored trio of distant moons. It orbits over 2 million miles from the planet. Iapetus is a fairly large moon, measuring about nine hundred miles across. One side of Iapetus is much darker than the other. Although the moon is almost

Titan, Saturn's largest moon (above left), and the haze that surrounds it.

entirely made of ice, scientists think the darker portions might be rock.

The smallest of Saturn's moons is also the farthest from the planet. With a diameter of just under 140 miles, tiny Phoebe orbits about 8 million miles from Saturn. The distance is so great that it takes over five hundred days for the moon to travel once around the planet. And, oddly, Phoebe orbits in the opposite direction of all the other moons.

Voyager 1 provides a wonderful view of Saturn and some of its moons.

Because it is so far from the planet, very little is known about Phoebe. Scientists believe that Phoebe may not have originally been involved in the formation process of Saturn and the other moons. Instead, Phoebe may be an **asteroid**—a large space rock—that was captured by the gravity of the newly forming planet. This theory has some merit, because unlike Saturn's other moons, Phoebe is composed mainly of rock instead of ice.

The Wonder of the Saturn System

As with its outermost moons, there is still much to learn about Saturn. In 1997 NASA cosponsored another mission to Saturn. This project is named after Gian Domenico Cassini, an early Italian astronomer who studied Saturn. The Cassini mission is designed to place an observation satellite in Saturn's orbit and send a probe into the atmosphere of Titan. Because of the spacecraft's weight and the great distance it has to travel, *Cassini* will not reach Saturn until 2004. But at that historic time, astronomers hope to learn much more about the mysterious planet.

The Cassini mission will increase astronomers' knowledge of Saturn, but until then the curious will continue to study the fascinating and beautiful photographs provided by the Voyager probes. With books and Internet websites devoted to the subject, these images are available to anyone interested in getting a closer look at Saturn and its many moons. Even the casual observer can

Using radar and other tools, *Cassini* will study Saturn and its moons.

step outside on a dark night and gaze at the bright light from the distant and wondrous ringed planet—as people have from the beginning of time.

Glossary

ammonia: A gas formed by combining the elements nitrogen and hydrogen. Ammonia is just one of the gases in Saturn's atmosphere.

asteroid: A chunk of rock that orbits the sun like a planet. Asteroids vary in size, and most exist in groups—called belts—that lie between Mars and Jupiter. Some scientists believe Saturn's outermost moon, Phoebe, may be an asteroid that Saturn's gravity pulled from its solar orbit near Jupiter.

comet: A heavenly body with a starlike center and a long luminous tail. Comets orbit the sun and sometimes crash into planets and moons.

diameter: The distance across the center of a circle. Since planets are circular when viewed from any side, the diameter is in line with the planet's equator.

gravity: The force of attraction. Gravity on Earth draws all matter to the ground. Saturn has no visible "ground," but gravity still pulls everything toward the planet's center.

helium: A common element. Helium is one of the principle gases in Saturn's atmosphere.

hydrogen: Hydrogen is a very plentiful element in the solar system. It is the most abundant gas in Saturn's atmosphere, and Saturn contains more hydrogen than any other planet. Hydrogen exists as a gas in Saturn's upper atmosphere and as a liquid in the lower atmosphere.

meteor: A heavenly body with a solid, rocky form. Large and small meteors travel through space and are sometimes set on collision courses with planets and moons. Saturn's moons, for example, have been hit many times by meteors, leaving visible impact craters on their surfaces.

methane: Another one of the gases present in Saturn's atmosphere. It is a compound of carbon and hydrogen.

NASA: The National Aeronautics and Space Administration. NASA is America's chief conductor of space missions. It has taken part in sending the Pioneer, Voyager, and Cassini missions to Mars, Jupiter, Saturn, and beyond.

orbit: The path of one heavenly body revolving around another. The revolving body is held in orbit partly by the gravity of the central body. This helps explain why an orbiting body continues to revolve around the central body instead of flying off into space.

phosphorous: An element that scientists believe may be in Saturn's atmosphere because of the colors displayed in the planet's cloud layers.

satellite: Although the term can refer to a man-made probe that orbits a planet or moon, a satellite may also be a heavenly body that does the same. In speaking of Saturn, astronomers usually use the word "satellite" to refer to the small rocklike moons that orbit the planet. This helps make a distinction between larger, round moons and smaller, irregularly shaped satellites. Technically, however, the terms "moon" and "satellite" are interchangeable.

solar system: The sun and the collection of planets, moons, and other smaller planetoid objects (such as asteroids) that revolve around it. Saturn is the sixth planet from the sun.

For Further Exploration

Don David and Ian Halliday, *Saturn.* New York: Facts On File, 1989. Part of the Planetary Exploration series. A good source of basic information about Saturn. The book is divided into short topics to make the information easy to find and understand. Pictures and illustrations are included.

Elaine Landau, *Saturn.* New York: Franklin Watts, 1999. An excellent overview of the planet for young readers. The book is filled with color photographs and illustrations as well as a fine narrative. Especially worthwhile is a section on the creation and goals of the Cassini space mission.

Patricia Lauber, *Journey to the Planets.* New York: Crown, 1993. An introductory guide to the planets and space exploration. From the landing on the moon to probes to other planets, this book covers what we know about heavenly bodies and how we know it.

NASA Kids (http://kids.msfc.nasa.gov). This is the space organization's website designed especially for kids. There is very basic information here on the planets and many fun projects to try. The site is updated regularly with reports on the best times to view astronomical events as well as upcoming NASA projects.

NASA's Cassini Project Website (www.jpl.nasa.gov/cassini). NASA's official website concerning the Cassini space mis sion. The site contains information on the construction of the spacecraft as well as what it is designed to accomplish. General information on Saturn can be found through various links.

Index

David M. Haugen edits books for Lucent Books and Greenhaven Press. He holds a master's degree in English literature and has also worked as a writer and instructor.

CARVER PUBLIC LIBRARY

3 2033 00083 9905